# Rare Earths:

## The Hidden Power of Critical Element

## The Materials that Control the Future

# Rare Earths:

## The Hidden Power of Critical Element

## The Materials that Control the Future

*(Illustrated, Refined & Edited)*

**By: DD Books**

DD BOOKS
*Publishers for New Generations*

DD BOOKS
*Publishers for New Generations*

## Copyright Notice for Readers



---

Publication Year: **2026** | Publisher: **DD Books**
ISBN: **9789372446623**

Edition: **First Edition, 2026** | Contact: ddsalesglobal@gmail.com

Book Name: **Rare Earths: The Hidden Power of Critical Element**
Author: **DD Books**
Price: **Rs 250/-**

# Table of content

# Reader's Note

There are certain forces that shape human civilization so completely that they become invisible. They do not announce themselves with spectacle, nor do they demand daily recognition, yet they quietly determine how societies function, how economies grow, and how individuals experience the world. Rare earth elements belong to this category of hidden forces. They reside beneath mountains, inside laboratories, and within the smallest components of modern devices, silently enabling the technologies that define contemporary life. Rare Earths: The Hidden Power of Critical Elements is an invitation to uncover these forces and to see the modern world with newly awakened eyes.

This book begins with a simple but profound idea: what we cannot see often matters most. We touch screens, hear engines, rely on networks, and trust machines to guide, diagnose, calculate, and communicate, yet rarely consider the materials that make such reliability possible. The elements explored in these pages are not household names, but their influence extends into nearly every corner of modern existence. By focusing on these substances, the book reveals a deeper layer of reality—one where chemistry intersects with culture, physics meets philosophy, and human ambition encounters natural limitation.

Readers approaching this book may come from many paths. Some may possess scientific backgrounds, others may be drawn by curiosity alone. No matter the starting point, the journey offered here is one of discovery rather than instruction. The goal is not to overwhelm with technical detail, but to illuminate with understanding. Scientific accuracy is carefully maintained, yet always placed within a broader narrative that emphasizes meaning, relevance, and human context.

Each chapter centers on a single element, but none stands alone. Together, they form a constellation of stories that reveal how rare earth elements function as the skeleton of modern innovation. Neodymium introduces the concept of invisible power, showing how extraordinarily strong magnets drive motion, store energy, and make digital communication possible. Dysprosium follows as a lesson in endurance, demonstrating how certain materials must perform under extreme conditions if advanced technology is to survive. Praseodymium opens a window into greener futures, illustrating how elemental choices shape environmental outcomes. Terbium highlights precision, Cerium and Lanthanum emphasize reliability, and Yttrium reveals the hidden architecture behind advanced systems.

Through these narratives, the book presents a quiet truth: progress is not created by a single breakthrough, but by countless small advantages accumulating over time. Rare earth elements provide those advantages. Their unique properties allow engineers to design machines that are smaller, lighter, stronger, faster, and more efficient than ever before. The cumulative effect of these improvements is nothing short of civilizational transformation.

Yet this book is not a celebration of technology alone. It is also an exploration of responsibility. Every benefit carries a cost. Mining alters landscapes. Processing generates waste. Global demand creates geopolitical tension. The story of rare earth elements is inseparable from questions about sustainability, equity, and long-term planning. By presenting these realities honestly, the book invites readers into a more mature understanding of progress—one that recognizes complexity rather than denying it.

What distinguishes this work is its insistence that science is a human story. Behind every discovery lies curiosity. Behind every refinement lies persistence. Behind every industrial application lies a chain of human effort stretching across continents and generations. The elements themselves may be ancient, but their modern significance is the product of human imagination. Readers may find themselves surprised by how emotionally resonant this subject becomes. There is something profoundly moving about realizing that atoms forged in cosmic events billions of years ago now participate in human communication, healing, and creativity. This awareness fosters a sense of connection that transcends time and scale. It reminds us that humanity does not stand apart from nature, but operates within it, shaping and being shaped in return.

The tone of the book is deliberately balanced. It does not indulge in technological utopianism, nor does it descend into pessimism. Instead, it presents innovation as a powerful tool—neither inherently good nor inherently harmful, but shaped by the intentions and choices of those who wield it. This perspective respects the reader's capacity for thoughtful judgment. Another defining feature of this book is its emphasis on interconnection. Rare earth elements reveal how deeply intertwined modern systems have become. A smartphone represents not just design and software, but mining operations, chemical processing, global logistics, political agreements, and environmental consequences. Understanding these connections fosters a more holistic view of the world. This holistic perspective has practical implications. When readers recognize the hidden complexity behind everyday objects, they may begin to approach consumption differently. Awareness can lead to appreciation, and appreciation can lead to responsibility. While the book does not prescribe specific actions, it quietly encourages mindfulness.

The reading experience offered here is meant to be immersive. The language flows with intention, allowing ideas to unfold gradually. Rather than racing toward conclusions, the narrative invites contemplation. This pace reflects the belief that deep understanding cannot be rushed. For readers who enjoy intellectual exploration, this book offers fertile ground. It connects chemistry to physics, engineering to environmental science, and technology to philosophy. These intersections create a rich tapestry of ideas that reward careful attention. At the same time, the book remains grounded. It consistently returns to real-world examples, demonstrating how rare earth elements shape devices, infrastructure, and industries that readers encounter daily. This balance between abstraction and application keeps the narrative accessible. Ultimately, Rare Earths: The Hidden Power of Critical Elements is about awareness. It seeks to expand the reader's mental map of reality. After reading, the world may appear subtly altered. A wind turbine may no longer be just a structure on the horizon, but a symbol of material ingenuity. A smartphone may feel less like a disposable object and more like a concentrated expression of scientific history. This shift in perception is perhaps the book's greatest achievement. It does not simply add information; it deepens understanding. If you choose to walk through these pages, you are choosing to engage with the hidden architecture of modern life. You are choosing to look beneath surfaces, to ask better questions, and to embrace complexity. That willingness—to remain curious in a complicated world—is itself a powerful form of intelligence. May this book reward your curiosity, challenge your assumptions, and remind you that even the smallest elements can carry immense significance.

# Publisher's Note

Every book arrives in the world carrying more than words. It carries intention, responsibility, and the quiet hope that it will reach readers who are ready to listen. Rare Earths: The Hidden Power of Critical Elements is one such book. From its earliest conception, it has stood as a work guided by respect—for science, for truth, for complexity, and for the intelligence of its audience. Our role as publisher has been to serve this vision faithfully while shaping an edition that speaks clearly and powerfully to modern readers.

We live in a time when information is abundant but understanding is scarce. Vast quantities of data circulate freely, yet much of it remains fragmented, simplified, or detached from meaningful context. In this environment, books that offer depth, coherence, and thoughtful perspective are more important than ever. This work does not seek to compete with the speed of headlines. Instead, it invites readers to slow down, to think, and to engage with ideas that unfold gradually and reward sustained attention. From the outset, we recognized that this book occupies a unique space. It is not a conventional science textbook, nor is it a purely narrative work of popular nonfiction. It exists at the intersection of rigorous explanation and reflective storytelling. Preserving this balance has been central to our editorial approach. The author's voice is marked by clarity, curiosity, and quiet authority. It does not announce itself with grand declarations, yet it carries a steady confidence rooted in careful research and deep contemplation. Our task was not to reshape this voice, but to refine it—smoothing transitions, enhancing flow, and ensuring that complex ideas are presented with precision and accessibility.

One of our guiding principles in preparing this edition has been fidelity. Fidelity to the author's intent. Fidelity to the integrity of the subject matter. Fidelity to the reader's experience. These three commitments often require careful negotiation. Simplifying too much risks losing depth. Preserving too much technical density risks alienating readers. The balance we have sought is one where sophistication and readability coexist.

Rare earth elements are, by their nature, complex. Their chemistry, physics, and industrial applications involve layers of specialized knowledge. Yet the author approaches these subjects with a humanistic sensibility, consistently connecting scientific facts to broader questions about society, ethics, and the future. Our editorial work has emphasized these connections, ensuring that each chapter feels not only informative, but meaningful.

We believe that books endure not because they explain everything, but because they illuminate something essential. This book illuminates the hidden foundations of modern life. It reveals how materials that most people never think about quietly shape the technologies they depend on every day. In doing so, it encourages a more conscious relationship with the material world.

Another important aspect of this edition is structural coherence. Each chapter focuses on a specific element, yet together they form a unified narrative arc. We have paid close attention to pacing, progression, and thematic continuity. Readers move from one element to the next not as if turning isolated pages, but as if advancing through stages of a larger exploration. Language plays a central role in this experience.

Scientific terminology is used with care, always supported by explanation and context. Metaphor and analogy are employed not to oversimplify, but to clarify. The result is prose that remains grounded in accuracy while inviting imaginative engagement.

We are particularly mindful that many readers approach science with a mixture of curiosity and apprehension. Past educational experiences may have left them feeling excluded from technical subjects. This book quietly challenges that barrier. It demonstrates that science is not an elite domain reserved for specialists, but a human endeavor that belongs to everyone.

Our editorial team has worked closely with this principle in mind. Sentences have been shaped for rhythm and clarity. Paragraphs have been structured to guide the reader through complex ideas step by step. The aim has been to create a reading experience that feels welcoming rather than intimidating.

At the same time, we have resisted the temptation to dilute the material. The book respects its readers. It assumes they are capable of thoughtful engagement. It does not avoid difficult questions or uncomfortable realities. Environmental consequences, geopolitical tensions, and ethical dilemmas are addressed with honesty. This candor is essential. A book about materials that shape the future must also confront the responsibilities that accompany such power.

As publishers, we are deeply aware that every book participates in a larger cultural conversation. Rare Earths: The Hidden Power of Critical Elements enters a world grappling with climate change, resource scarcity, and rapid technological transformation. In this context, the book serves not as a manifesto, but as a lens. It offers readers a way to see these challenges through the perspective of materials science, revealing connections that might otherwise remain hidden. We also view this edition as part of a long tradition of scientific writing that seeks to bridge disciplines. The greatest works in this tradition do more than explain experiments or describe discoveries. They ask what those discoveries mean for humanity. They invite readers to consider not only how things work, but why they matter.

This book belongs firmly within that lineage. Our production choices reflect this philosophy. The design emphasizes readability and visual calm. The layout encourages sustained reading. The typography supports clarity. These details may seem small, but they shape the reader's experience in profound ways.

We are grateful for the opportunity to present this work to a global audience. We believe it will resonate with students encountering materials science for the first time, with professionals seeking broader perspective, and with general readers who simply wish to understand the hidden forces shaping their world.

Most of all, we hope this book will cultivate a sense of thoughtful curiosity. Curiosity that does not end with the final page, but continues into daily life. Curiosity that prompts readers to look at familiar objects with new awareness. Curiosity that recognizes that the modern world is not magic, but the product of countless human and natural processes working together. In publishing this edition, we reaffirm our commitment to producing books that matter. Books that challenge, illuminate, and endure. Books that respect both their subjects and their readers. It is our honor to place Rare Earths: The Hidden Power of Critical Elements into your hands.

# Introduction

The modern world presents itself as a miracle of speed, precision, and connectivity. Screens respond to the slightest touch. Vehicles glide almost silently along roads. Information moves across continents in fractions of a second. Medical instruments peer inside the human body with astonishing clarity. All of this appears seamless, almost effortless. Yet beneath this smooth surface lies a hidden reality—an intricate architecture of matter, energy, and design that makes such wonders possible. This book begins where most stories of technology end: at the elemental level, in the quiet domain of atoms whose influence far exceeds their size.

Rare earth elements occupy a strange and fascinating position in human history. They are neither truly rare nor widely understood. For centuries, they existed as obscure curiosities in mineral collections and laboratory shelves. Today, they stand at the heart of some of the most advanced technologies ever created. This transformation did not happen overnight. It is the result of patient experimentation, cumulative knowledge, and the persistent human drive to push beyond perceived limits.

The story told in these pages is not a single narrative with a single hero. It is a mosaic of discoveries, failures, breakthroughs, and unexpected connections. It is a story in which chemistry meets engineering, physics meets environmental science, and human ambition meets natural constraint. It is also, fundamentally, a story about power—not the loud, visible power of machines and markets, but the quiet power embedded in the structure of matter itself.

Consider the objects that surround you as you read these words. Somewhere within them are atoms of neodymium, dysprosium, praseodymium, terbium, cerium, lanthanum, and yttrium. These atoms perform their roles silently, faithfully, without recognition. They convert electrical energy into motion, stabilize materials under extreme heat, emit precise colors of light, and enable digital memory to function. Without them, many of the technologies that define contemporary life would simply not exist.

Yet most people have never heard their names. This book seeks to give voice to these silent contributors. Not in the form of dry lists or abstract formulas, but as living participants in a human story of ingenuity and transformation. Each element becomes a lens through which to view a different aspect of modern civilization.

Neodymium introduces us to the concept of invisible force. Its magnetic strength underpins countless devices, from headphones to wind turbines. Dysprosium reveals the importance of endurance, showing how materials must survive brutal conditions if advanced machines are to function reliably. Praseodymium opens a conversation about sustainability and the pursuit of cleaner technologies. Terbium brings attention to precision and the science of light. Cerium and lanthanum highlight the value of reliability and adaptability. Yttrium demonstrates how certain elements quietly enable entire classes of advanced applications.

Together, these stories form a portrait of a world built not only by grand ideas, but by microscopic advantages. The emotional core of this book lies in a simple realization: humanity's greatest achievements are inseparable from the natural world. We do not create materials from nothing. We discover them. We learn their properties. We find ways to collaborate with nature at the atomic level. This partnership is both empowering and humbling.

There is wonder in this relationship. The atoms that now drive electric motors were formed in stellar processes long before Earth existed. They traveled through cosmic history, became part of this planet, and eventually entered human hands. When you hold a device containing rare earth elements, you are, in a sense, holding a fragment of the universe's ancient past shaped into a tool for the present. At the same time, this relationship carries responsibility. Extracting and processing rare earth elements is not without cost. Landscapes are altered. Ecosystems are affected. Communities are impacted. Supply chains become entangled with political and economic power. These realities form an essential part of the story.

This book does not present easy answers. Instead, it presents honest questions. How can societies balance technological advancement with environmental stewardship? How can innovation serve humanity without deepening inequality? How can we build systems that are not only powerful, but just and sustainable?

By exploring rare earth elements, the book enters these broader conversations through a tangible and concrete doorway. Materials become a way to think about values.

One of the defining qualities of the modern age is acceleration. New devices appear constantly. Software updates reshape experiences overnight. Entire industries rise and fall within a generation. Amid this rapid change, it is easy to forget that every innovation rests on a foundation of physical reality. Rare earth elements remind us of that foundation.

They also remind us that progress is cumulative. No single discovery created the digital age. No single scientist or engineer made it inevitable. Progress emerges from networks of effort, often spanning decades and continents. This book honors that collective endeavor.

The narrative voice of this work is intentionally reflective. It does not rush toward conclusions. It lingers on implications. It invites readers to pause and consider. In doing so, it offers a different kind of reading experience—one that values depth over speed. You will encounter science in these pages, but you will also encounter philosophy. You will encounter industry, but also ethics. You will encounter technical explanation, but also moments of wonder. This blending reflects the belief that knowledge is richest when it crosses boundaries.

For some readers, this book may change the way technology feels. Devices may no longer seem like sealed black boxes, but like intricate collaborations between human design and elemental behavior. For others, it may inspire a deeper interest in materials science, engineering, or environmental studies. For many, it may simply foster a quiet appreciation for the complexity of the world.

This introduction is not meant to summarize everything you will find ahead. It is meant to prepare you for a journey of attention. Attention to the small Attention to the unseen. Attention to the forces that quietly shape modern existence. As you move forward, you will begin to notice patterns. How strength often hides behind subtlety. How reliability matters as much as brilliance. How progress depends not only on breakthroughs, but on balance. You will also encounter a recurring theme: the idea that the future is not predetermined. The materials exist. The knowledge exists. But how they are used remains a human choice.

# Chapter-1.
# Neodymium:
# The Invisible Engine of Modern Innovation

*Hidden within the machines that power modern life lies neodymium—a quiet element driving electric mobility, renewable energy, and advanced technology. This chapter introduces neodymium as the invisible engine of innovation, revealing how a single material shapes global progress, geopolitical strategy, and sustainability. Understanding its role equips readers to grasp how the future is built, powered, and responsibly sustained.*

In the age of rapid technological transformation, where machines think, move, and respond with unprecedented precision, there lies a silent force that powers this revolution—a metal not spoken about in daily conversations, yet woven into the very spine of modern civilization.

This chapter unveils the remarkable story of neodymium, a rare-earth element whose strength lies not in its visibility but in its invisible grip on the future. From the whir of an electric vehicle to the hum of a wind turbine, neodymium plays a pivotal role. Understanding this element is not merely about chemistry or mining; it's about comprehending how the future runs, drives, flies, and sustains itself. For young minds stepping into a world governed by smart technology and sustainable choices, knowing neodymium is knowing tomorrow.

## What is Neodymium and Why it Matters

Neodymium is a lanthanide, one of the light rare-earth elements that occurs in nature bonded with other minerals, hidden in the crust of the Earth and extracted through labor-intensive processes. Though not rare in terms of abundance—it ranks about 27th among elements in the Earth's crust—it is rarely found in concentrated forms, making its extraction challenging. But what elevates neodymium from a common mineral to a strategic marvel is its exceptional magnetic properties. When alloyed with iron and boron, neodymium becomes the heart of NdFeB magnets, the strongest known permanent magnets on Earth. These magnets are not only compact and durable, but they also retain their magnetic field far longer and with more intensity than any conventional counterpart.

The brilliance of neodymium magnets lies in their ability to deliver immense power in minimal space. A small neodymium magnet can perform the task of a traditional magnet several times its size.

This property alone revolutionized the miniaturization of electronic devices. Smartphones, laptops, earbuds, and hard drives owe their sleek designs to neodymium's power density. But beyond consumer electronics, the true genius of neodymium is revealed in sectors where efficiency, weight, and performance converge into a single imperative.

Electric vehicles, the vanguard of clean transportation, depend on lightweight, high-torque motors made possible by neodymium-based magnets.

These motors convert electric energy into mechanical power with minimal loss, giving EVs their hallmark smoothness and acceleration. Similarly, wind turbines harness wind energy more efficiently with neodymium magnets embedded in their generators. Their ability to produce continuous rotation and strong torque even at low wind speeds makes them ideal for renewable power generation.

What makes neodymium irreplaceable is not just its strength, but the fact that few materials can replicate its magnetic performance without significant trade-offs in size, weight, and energy efficiency. While alternatives are being explored, they either fall short in strength or dramatically increase cost and complexity. This element, therefore, represents a bottleneck in the supply chain—its scarcity not in quantity, but in accessibility, refinement, and geopolitical control.

A substantial portion of global neodymium production is controlled by a handful of nations, with China refining and processing the vast majority. This concentration of supply has raised strategic concerns among governments and industries worldwide. Neodymium is no longer just a resource—it is a leverage point in international relations, a matter of energy independence, and a key to economic resilience. Countries are now racing to explore domestic reserves, invest in recycling technologies, and build supply chains that can shield them from geopolitical uncertainties.

At the frontier of defense and aerospace, neodymium plays another crucial role. Its magnets are integrated into guided missile systems, stealth aircraft actuators, and secure communication systems. In outer space, where every gram matters and failure is not an option, neodymium's compact power delivers reliability. It is the unsung material behind high-precision control surfaces, satellite propulsion systems, and deep-space communication equipment.

Yet, with all its potential, neodymium poses challenges that extend into environmental ethics. The process of mining and refining rare-earth elements often involves the use of hazardous chemicals, generates radioactive waste, and disrupts ecosystems. This dilemma poses a moral question to young engineers, scientists, and innovators: how do we pursue high technology without compromising planetary well-being? The future of neodymium lies not just in extraction, but in sustainable stewardship. Recycling discarded electronics, developing cleaner extraction methods, and substituting harmful processes with greener alternatives will define the next chapter of its story.

To understand neodymium is to understand how a single element can bridge the gap between raw geology and human innovation. It sits at the core of a new industrial age—not one fueled by coal and steel, but by rare elements and information. Its influence stretches across disciplines: physics, material science, energy policy, global economics, and environmental justice. For youth aspiring to shape the future, neodymium is more than a chemical—it is a case study in complexity, potential, and responsibility.

By learning about neodymium, young readers glimpse the intricate machinery that runs behind the scenes of modern life. They begin to see the hidden threads that link sustainability with science, technology with ethics, and innovation with global collaboration. In mastering the narrative of this single element, they gain insights into a future that they themselves must power, design, and protect.

## Explosive Demand Across Sectors

In the ever-accelerating race toward a cleaner, smarter, and more connected world, certain materials quietly rise from obscurity to necessity. Neodymium, a rare-earth element once confined to the curiosity of chemists, now finds itself at the epicenter of this transformation. This chapter offers young readers a deep understanding of how neodymium's magnetic prowess is driving revolutions across diverse sectors—from mobility to medicine, from renewable energy to intelligent machines. To understand where the world is going, one must first understand where its demand is growing.

The surge in electric vehicle adoption has not only redefined transportation—it has redefined the materials that matter. At the heart of nearly every high-performance electric motor lies a neodymium-iron-boron (NdFeB) magnet. Compact, durable, and powerful, these magnets convert electric energy into mechanical motion with remarkable efficiency. Each electric vehicle contains roughly 1 to 2 kilograms of these magnets, depending on the number of motors and the torque requirements of the model. A widely used vehicle such as the Tesla Model Y, with its dual-motor configuration, utilizes about 500 grams of NdFeB magnets in each motor. Multiplied across the global production of electric vehicles—13.6 million units sold in 2023 and an anticipated 17 million in 2024—the numbers become staggering. Electric vehicles now account for approximately 35 to 40 percent of all neodymium magnet usage worldwide, and this figure is climbing with every passing year. These magnets enable the compact, high-efficiency motors that give EVs their seamless acceleration, long range, and energy savings—qualities that are non-negotiable in the shift away from fossil fuels.

Parallel to the revolution on roads, the sky and earth themselves are being tapped for energy. Wind turbines—especially direct-drive systems used in both offshore and onshore installations—rely heavily on neodymium magnets. These turbines convert wind energy into electricity using permanent magnets embedded within their generators, eliminating the need for complex gearboxes and improving long-term reliability. Depending on design, each megawatt of turbine capacity can require between 200 to 600 kilograms of neodymium magnets. A single 3-megawatt wind turbine may therefore contain close to 600 kilograms of this critical material. With global wind power installations surpassing 75 gigawatts in new capacity in 2023 and projections pointing toward steady growth—particularly in offshore wind—the demand for neodymium in this sector alone is expanding rapidly. What makes this even more significant is that the magnets in wind turbines operate continuously, under stress, in harsh environments—making their quality and performance a matter of both engineering and economics.

Yet the reach of neodymium extends far beyond the massive and the industrial. In the world of personal technology, where compactness and performance are paramount, neodymium magnets enable many of the conveniences young people interact with every day. Inside wireless earbuds, headphones, mobile phones, and hard drives, one finds slivers of magnetic material—often less than one gram in mass—responsible for precise sound, vibration, and movement.

Without these powerful magnets, the miniaturization of electronics as we know it would be impossible. The same principle extends into emerging fields like robotics. As intelligent machines increasingly enter factories, hospitals, and even homes, neodymium magnets become essential for their motion. Servo motors, which control the joints and limbs of robotic arms, rely on these magnets for speed, accuracy, and torque.

A single robotic unit may incorporate 20 to 30 magnets of varying sizes, with each one contributing to its balance, reach, and precision. This is not merely a matter of convenience—it is the mechanical language of the Fourth Industrial Revolution.

In the realm of health care and defense, where stakes are immeasurably high, neodymium again plays a vital role. Magnetic Resonance Imaging (MRI) machines, especially those operating at 1.5 or 3 Tesla, depend on high-strength magnets for the generation of detailed internal images. These magnets must be stable, powerful, and durable, often operating in cryogenic conditions. Neodymium enables this performance with greater efficiency and lower energy consumption than many traditional alternatives. Furthermore, as portable imaging devices grow in importance for emergency and rural medicine, neodymium's compact power makes real-time diagnostics more accessible and mobile.

On the battlefield and in national security domains, precision and power must go hand in hand. Neodymium magnets are embedded in the motors and sensors of advanced missiles, sonar arrays, unmanned aerial vehicles (UAVs), and stealth aircraft actuators.

Their presence allows for faster reaction times, higher maneuverability, and minimal detectability. In these applications, performance is not simply measured in watts or grams—it is measured in lives protected, missions succeeded, and borders secured.

As sectors diversify and digitize, the range of applications for neodymium continues to expand. What unites all these uses—whether in a child's toy or a $100 million turbine—is the demand for compact, reliable magnetic force.

The versatility of neodymium magnets stems from a singular property: the ability to deliver extreme strength in minimal space. This allows engineers and designers to innovate without compromising on size or efficiency. But this very advantage also contributes to growing global pressure on supply, pricing, and sourcing ethics—topics explored further in the chapters ahead. For young innovators and thinkers, understanding how a single material like neodymium shapes such a vast array of industries is both eye-opening and empowering. It reveals how science and society interlock, how materials determine possibilities, and how the demands of technology ripple through the fabric of the planet. As tomorrow's engineers, policymakers, and creators, it is this awareness that will allow them to build a world not just of speed and strength, but of sustainability and wisdom.

## Production and the Global Supply Chain

Behind every innovation that uses neodymium magnets—whether it's an electric vehicle, a wind turbine, or a piece of cutting-edge medical imaging equipment—lies a complex, tightly knit global supply chain. Understanding where neodymium comes from and how it moves from ore to high-performance magnet is not just a matter of industry knowledge; it is a lesson in economics, politics, environmental responsibility, and global cooperation.

This chapter opens a window into the hidden infrastructure that fuels the technological future, inviting young readers to grasp not only how things are made, but also the power dynamics behind who controls what. The modern supply of neodymium and its related compounds is neither equal nor decentralized. In fact, it is one of the most geopolitically concentrated mineral markets on Earth. China currently dominates nearly every stage of the rare-earth production pipeline. It mines approximately 60 to 70 percent of the world's rare-earth ore, but more critically, it controls close to 90 percent of the global capacity to refine, separate, and fabricate those raw materials into usable forms—especially magnets. This kind of concentration grants significant leverage to one nation over a resource that powers electric mobility, renewable energy, and even modern warfare systems.

This dominance is not an accident; it is the result of decades of strategic investment. China has invested heavily in refining technologies, workforce specialization, and vertically integrated supply chains that can convert mined ore into ready-to-use materials at scale. Over time, other countries scaled back their own mining operations due to cost, environmental hurdles, and a perceived lack of necessity. Now, as demand skyrockets and the world realizes the critical nature of rare-earth elements, many nations are scrambling to catch up. The problem is not just that supply is concentrated; it's that the upstream supply—the first stages of mining and ore processing—are particularly vulnerable to disruption. Natural disasters, trade conflicts, diplomatic strains, or internal policy shifts in China could all suddenly squeeze the global flow of neodymium. For industries that rely on just-in-time logistics and tightly integrated global sourcing, even a small interruption in magnet production can ripple across entire sectors—from slowing EV production lines to delaying wind farm installations.

This geopolitical reality has triggered a wave of responses. Countries and companies are now racing to develop alternate sources of neodymium and other rare-earth elements, aiming to reduce dependency and secure long-term access. Australia has emerged as a leading player in this space. Its Mt Weld mine, operated by Lynas Corporation, is one of the richest rare-earth deposits outside China. In early 2025, Lynas announced a 16.7% increase in its neodymium-praseodymium (NdPr) oxide output, with further expansion underway. This boost not only adds volume to the market but also signals growing maturity in non-Chinese supply chains.

In the United States, a significant development has occurred at Mountain Pass, California. Once a dormant operation, the site has been revived by MP Materials, which is now vertically reintegrating by resuming in-house production of NdFeB magnets. Their new processing facility in Texas began pilot operations in early 2025, with commercial shipments expected soon after. This shift marks a critical milestone for American industrial strategy: transforming from raw ore exporter to value-added magnet producer.

Across the Atlantic, the United Kingdom is positioning itself as a major player in rare-earth refining.

Pensana's Salt End facility, under development in Yorkshire, plans to supply about 5 percent of the global NdPr oxide demand by 2025. Importantly, the plant sources its ore from Angola, creating a diversified supply chain that spans multiple continents. Pensana is also focused on environmentally responsible practices, aiming to set new standards for clean rare-earth processing in Europe.

The reshaping of the global market doesn't stop with individual mines or factories. A new era of industrial alliances and technological cooperation is forming to challenge China's monopoly.

Companies such as e-VAC Magnetics in the United States and Neo Performance Materials, with operations in Canada and Estonia, are scaling up their capabilities. Some are focusing on refining and separation; others are diving into downstream magnet production. Mergers, joint ventures, and strategic government partnerships are becoming common as the world builds resilience into its rare-earth infrastructure.

These new players are not merely competing with China—they are diversifying the model.

Rather than centralizing the entire supply chain in one region, the emerging approach favors regional integration. One country may mine, another may refine, and a third may assemble the final components. While this structure is more complex, it is also more stable, providing a buffer against political risk and logistical bottlenecks.

This growing diversification is not without its challenges. Mining rare-earth elements remains environmentally intensive, and building refineries from scratch involves years of permitting, construction, and testing. Moreover, the market must remain economically viable; rare-earth prices are highly volatile, and without long-term demand guarantees, many projects struggle to attract investment. Yet this transitional moment offers profound opportunities—especially for the next generation. As global attention shifts toward sustainability, circularity, and strategic resource planning, young thinkers have a chance to lead. Whether as engineers optimizing separation processes, entrepreneurs innovating recycling methods, or diplomats forging trade alliances, they can redefine how the world sources and shares its most critical materials.

Understanding the production and supply chain of neodymium is more than academic.

It teaches the delicate balance between innovation and dependency, between abundance and control. It is a story of not just what we build, but how we sustain it.

And in this story, the next chapter will be written by those who understand that materials, like ideas, must be cultivated wisely, traded justly, and used responsibly.

## Challenges and Environmental Concerns

Every technological breakthrough carries with it a shadow—an underside that, if ignored, may undermine the very progress it claims to bring. Neodymium has rapidly become a keystone of the clean energy and high-tech revolutions, but its extraction, processing, and supply chain raise significant economic, environmental, and ethical concerns. This chapter unpacks those hidden costs, encouraging young readers to see beyond the gleam of innovation and into the systems that must be balanced, reformed, and reimagined if sustainable development is to be truly achieved. Understanding these challenges is the first step toward becoming a responsible innovator, scientist, or policymaker in the rare-earth era.

The market dynamics of neodymium are marked by a volatility rarely seen in more mature commodity sectors. Because its supply is concentrated in a handful of regions—particularly China—and demand is growing across multiple high-growth industries, prices for neodymium and its compound forms are highly sensitive to political decisions, trade restrictions, and even rumors of supply disruptions. A diplomatic tension or export ban can trigger sudden spikes, making long-term planning difficult for manufacturers of electric vehicles, wind turbines, or robotics. This instability introduces significant economic risk, not only for companies but also for entire national supply strategies. For countries investing in rare-earth-dependent infrastructure, pricing uncertainty is more than a budgetary problem—it is a question of strategic security.

But economic volatility is only one part of the story. The environmental footprint of neodymium mining and refining is deeply concerning and increasingly unacceptable under modern sustainability standards. Most rare-earth ores are chemically complex, and separating out neodymium requires the use of harsh acids and solvents. The refining process generates large volumes of toxic waste, including radioactive thorium and uranium byproducts, which must be handled with extreme caution. In several mining regions, waste storage has led to serious soil and water contamination, threatening both ecosystems and human health. As global awareness grows, so too does resistance from local communities and environmental activists. What was once tolerated as a cost of progress is now being questioned at every level—scientifically, ethically, and legally.

These concerns have prompted governments and companies to explore stricter regulations, more transparent reporting, and environmentally friendlier methods of extraction. Yet adopting new techniques is far from simple. Traditional mining remains the most commercially viable approach, and transitioning to cleaner alternatives requires large upfront investments, long testing cycles, and reliable technological breakthroughs. However, there are promising developments underway. Advanced separation techniques like high-flux solvent liquid membranes (HFSLM) and ionic liquid extractions are being researched for their ability to reduce chemical use and waste. These emerging methods aim to minimize environmental damage while improving yield and precision in isolating rare-earth elements like neodymium.

Perhaps the most transformative approach lies not in extraction, but in the concept of circularity—rethinking waste as a resource. Recycling end-of-life electronics, electric motors, and magnets could offer a secondary supply stream with far lower environmental impact. Currently, only about 1 percent of neodymium in circulation is recycled, largely due to technical and economic barriers. Magnets are often embedded deep within devices, making them hard to retrieve without damaging their magnetic properties. However, initiatives such as advanced robotic disassembly, chemical recycling, and magnet-to-magnet regeneration are showing promise. The vision is clear: create a closed-loop supply chain in which today's discarded phone or wind turbine becomes tomorrow's new magnet, without ever digging another trench in the earth.

Programs like these require both innovation and infrastructure. Universities and startups are already pioneering magnet recycling systems that can scale, but without government support and consumer awareness, progress remains slow. For young minds entering these fields, the challenge is thrilling. This is not just about engineering efficiency or reducing costs—it is about redesigning the material economy itself. The next generation of scientists and technologists must not only create new devices, but also reimagine how those devices are made, used, and reborn.

At the heart of this challenge lies a philosophical question: can the tools of sustainability be built sustainably? The story of neodymium offers no easy answers, but it offers a powerful lesson. Technologies that solve one problem may create others, and true progress means thinking holistically. It means understanding that every magnet, every circuit, every innovation carries within it a set of invisible trade-offs—economic, ecological, and human.

This awareness does not signal defeat, but empowerment. By facing these challenges head-on—by studying them, discussing them, and designing around them—young innovators can lead a new era of ethical, clean, and responsible technology. In doing so, they can ensure that the age of rare earths does not become one of rare wisdom. Instead, it can become a chapter in human history where intelligence, integrity, and ingenuity finally worked as one.

## Outlook and Strategic Implications

As neodymium becomes increasingly vital to the machinery of modern civilization, its future looms large on the global stage—not only in terms of supply and demand, but in how entire industries, governments, and generations will adapt to its limitations and possibilities. This chapter invites young minds to think ahead: to imagine a world shaped by material scarcity, technological evolution, and geopolitical complexity. Understanding the strategic outlook for neodymium is more than an academic exercise; it is a necessary compass for navigating the future of sustainability, innovation, and economic resilience.

The demand trajectory for neodymium is both staggering and sobering. As the global transition toward electrification accelerates—driven by electric vehicles, renewable energy systems, robotics, aerospace, and digital infrastructure—the appetite for neodymium is expected to rise by nearly 48 percent by mid-century.

But the real stress point may arrive far sooner. Forecasts indicate that by 2030, global demand for neodymium and related rare earths could reach approximately 850,000 metric tons. If current production rates and technological strategies remain unchanged, the world may face a shortfall of nearly 250,000 tons—an alarming gap with far-reaching consequences.

Bridging that divide requires a multi-pronged strategy, none of which can succeed in isolation. The most immediate and impactful opportunity lies in recycling.

At present, less than 1 percent of neodymium is reclaimed from used products. This is not due to lack of material, but rather the complexity of extraction. NdFeB magnets are often encased deep within electronics and machines, and retrieving them without degrading their properties is a significant engineering challenge. However, to meet future needs sustainably, the recycling rate must rise dramatically—up to 15 percent by 2030. Achieving this will demand not just better technology, but also better systems for product design, material tracking, and consumer awareness. Design for disassembly, automated separation technologies, and chemical regeneration of magnets are all areas ripe for innovation. These advances would allow used magnets to reenter the supply chain with minimal performance loss and environmental impact. But recycling alone will not be enough. Even the most efficient circular systems require complementary inputs and alternatives.

This brings us to substitution—a field of rapid exploration and cautious optimism. Scientists and material engineers are investigating alternative magnetic materials that could reduce or eliminate neodymium dependence. One of the most promising candidates is tetrataenite, a naturally occurring iron-nickel alloy that mimics the magnetic behavior of rare-earth magnets without the same geopolitical and ecological burdens. Synthetic tetrataenite has recently become producible in laboratory settings, and although commercialization remains a challenge, its potential to replace or supplement neodymium in motors and generators is real.

Other substitution strategies focus on enhancing the performance of weaker magnets like ferrites or incorporating cerium, another rare earth that is more abundant and less environmentally taxing. Additionally, there is growing interest in rare-earth-free motor designs that use advanced electromagnetism and structural engineering to match the torque and efficiency of traditional systems. These innovations are not yet mainstream, but they represent crucial contingency plans in a world where material bottlenecks could stall entire sectors. Alongside recycling and substitution is the third and most traditional lever: expanding supply. Around the globe, new mines are being planned, old ones reopened, and unexplored reserves assessed. Countries once passive in rare-earth mining—such as Vietnam, Brazil, and various African nations—are now exploring partnerships, infrastructure investments, and policy reforms to enter the market. Processing capacity, historically centralized in China, is beginning to decentralize through the development of facilities in the U.S., Europe, Australia, and parts of Southeast Asia.

This diversification is not only about physical volume—it is about strategic balance. Governments and defense agencies are beginning to treat rare-earth elements as matters of national security. Funding programs, stockpiles, and private-public partnerships are emerging to stabilize supply, encourage innovation, and reduce vulnerability. Nations are realizing that material resilience is as vital as energy independence or digital infrastructure. This reframing of neodymium—from commodity to strategic asset—will shape future trade policies, research agendas, and industrial strategies. For the youth poised to inherit this material-intensive world, the strategic outlook for neodymium is both a warning and a roadmap. It demonstrates the necessity of foresight, adaptability, and interdisciplinary thinking. Material science must join hands with economics, environmental science, and political strategy. The future will not be built by isolated innovations, but by integrated systems that balance performance with responsibility, growth with sustainability, and ambition with realism.

In the end, neodymium is more than a magnet—it is a mirror. It reflects the promises and perils of our technological age. The question is not just whether we can find enough of it, but whether we can learn from it: to think smarter, build cleaner, and plan longer. The solutions will not arrive fully formed, but they will come—from the workshops, labs, and minds of the next generation ready to shape a future in which power is not only generated, but also wisely managed.

**End of the Chapter**

# Chapter-2.
# Dysprosium: The Silent Strength Behind High-Temperature Innovation

*Rarely noticed yet critically indispensable, dysprosium is the element that allows modern magnets to survive extreme heat and stress. This chapter reveals how dysprosium quietly strengthens electric vehicles, renewable energy systems, and advanced defense technologies. By understanding its unique role, readers gain insight into the material limits of innovation—and the strategic challenges of building a high-performance, sustainable future.*

In a world increasingly dependent on magnetic technologies—especially those powering electric mobility, renewable energy, defense systems, and precision electronics—certain elements, though obscure to the public, stand at the very heart of this transformation. Dysprosium is one of them.

Rare, powerful, and often invisible in everyday discourse, this element holds the key to maintaining magnetic performance under extreme conditions.

For students and young thinkers eager to understand how material science shapes the modern world, Dysprosium offers a compelling example of how a single element can define the limits—and possibilities—of future technology.

## What Dysprosium is and Why it

Dysprosium, symbolized as Dy with atomic number 66, belongs to the lanthanide series of rare earth elements. Its name, derived from the Greek word "dysprositos," meaning "hard to get," remains true to its character. Though it is not the rarest of the rare earths in terms of crustal abundance, its isolation and refining are challenging, making its availability limited and its value high. The true power of dysprosium lies not in its obscurity, but in its magnetic resilience. When added in small percentages to neodymium-iron-boron (NdFeB) magnets, it drastically improves their resistance to demagnetization at high temperatures. This property is not just a scientific curiosity—it is a technological necessity. Without dysprosium, many advanced magnetic systems would falter in the heat, literally and figuratively.

This singular trait makes dysprosium indispensable in the global push for electrification, especially in technologies that demand sustained performance under thermal stress. Electric vehicles are a prime example. Premium EV motors operate under high loads and elevated temperatures, often pushing the limits of conventional magnet stability. To avoid weakening or failure, manufacturers integrate dysprosium—typically 2 to 4 percent by weight—into NdFeB magnets used in the motor's rotor assembly. While that might sound like a small amount, in absolute terms, each vehicle may require 40 to 60 grams of dysprosium per motor. When multiplied by millions of vehicles, the demand becomes substantial. Today, the EV industry alone accounts for approximately 15 to 20 percent of the world's dysprosium consumption, and that figure is growing with every advancement in electric mobility.

The wind energy sector adds yet another layer of demand. In direct-drive wind turbines—especially offshore installations where maintenance is difficult and operational efficiency is crucial—permanent magnets must remain stable across a wide range of environmental conditions. Here too, dysprosium plays a quiet yet critical role. An 8-megawatt turbine may use 100 to 200 kilograms of dysprosium to maintain magnet strength and reliability over decades of performance. Without it, the efficiency of power conversion would drop, turbine downtime would rise, and the return on investment in renewable infrastructure would falter. In short, dysprosium ensures that clean energy isn't just powerful—it's dependable.

Beyond clean energy and mobility, dysprosium extends its influence into the compact, precise, and personal. In smartphones, laptops, and tablets, space is limited and heat is inevitable. Small Dy-enhanced magnets allow for miniaturization without compromising reliability. Whether it's a speaker that never buzzes under strain, a fan motor that spins without warping, or a camera lens that adjusts with silent precision, dysprosium makes these miracles of design possible. The same logic applies to robotics and industrial automation, where servo motors must respond with millisecond accuracy while operating continuously. Here, dysprosium allows machines to mimic the subtlety of human movement while enduring mechanical stress far beyond human tolerance.

In the defense and aerospace sectors, the stakes are even higher. Advanced missiles, satellites, sensors, and unmanned aerial vehicles (UAVs) operate in environments of extreme heat, speed, and vibration. Failure is not an option.

The magnetic systems used in these applications rely heavily on dysprosium to maintain alignment, guidance, and propulsion. It is this element that makes stealthy drone propulsion quiet and efficient, missile navigation systems precise, and satellite communications uninterrupted by temperature variations in space. While the quantities used may be small, the strategic implications are immense. Then there are applications that go deeper—literally and metaphorically. In the nuclear energy sector, dysprosium is used in control rods due to its strong neutron absorption capabilities. These rods regulate the fission process within reactors, ensuring safe and stable energy generation.

The importance of this function cannot be overstated, as it directly governs the safety protocols of nuclear power plants. In scientific and industrial research, dysprosium is found in specialized lasers, especially those operating in the infrared spectrum. These are used in spectroscopy, telecommunications, and certain military targeting systems.

Its role in high-density data storage also continues to evolve, particularly in technologies that must remain stable over time and under pressure. What binds all of these applications together—mobility, energy, defense, electronics, and data—is the fact that none could function at their current levels without dysprosium. It is not the most abundant, nor the most famous, but it is among the most functionally irreplaceable. And therein lies both its strength and vulnerability.

As demand rises and supply remains constrained, dysprosium has become a point of strategic concern for nations, industries, and researchers. Its unique properties are difficult to replicate, making the search for substitutes or recycling methods both urgent and ongoing.

For young scientists, engineers, and future policymakers, dysprosium is a lesson in material dependence. It teaches that the smallest inputs can define the largest outcomes, and that the true value of an element is not in its weight or visibility, but in its utility and irreplaceability.

To build a better world, one must know not just how things work—but what they're made of, and what those materials demand in return. Dysprosium, quiet and powerful, is one such element. It may be hard to get, but it is even harder to replace.

## Production & Supply Landscape of Dysprosium

The production of dysprosium is overwhelmingly dominated by a single country—China. With approximately 98 percent of the world's annual output, China's grip on dysprosium is unmatched by any other resource in the critical minerals sector. Most of this output comes from ion-adsorption clays in southern China, which are easier to mine but environmentally delicate.

Unlike traditional hard rock mining, these clays are leached using ammonium sulfate and other chemicals to extract rare earths, including dysprosium. This method has allowed for relatively low-cost and high-volume production, yet it comes with ecological costs—soil degradation, water contamination, and public resistance in surrounding regions.

China's control extends beyond mining. Through vertical integration, it manages nearly the entire supply chain—from ore extraction to refining, alloy manufacturing, and magnet production. This level of centralization allows China not only to dominate global pricing but also to strategically modulate supply, influencing industries across the globe. For emerging economies and technological manufacturers, this creates a delicate balance: they depend on a material over which they have little control.

The global community is not unaware of this imbalance. Alternative sources have begun to emerge, though at a slow and cautious pace. Myanmar, for instance, contributed around 31 percent of dysprosium supply in 2021 through similar clay deposits near its southern border. However, operations there are subject to both political instability and environmental scrutiny, which makes long-term reliability questionable. Australia is leading non-Chinese production with sites like Browns Range, which offers a hard-rock source of heavy rare earths. Though promising, such mines are capital-intensive and face regulatory hurdles.

Other countries—such as Brazil, India, and the United States—are developing their capabilities, but remain far behind in terms of scale, refining technology, and supply chain infrastructure. These regions often suffer from fragmented policy, lack of long-term investment, or insufficient geological data. For example, India possesses considerable rare-earth potential, but limited domestic refining capacity and export restrictions have slowed progress.

An emerging frontier in dysprosium supply is recycling—still in its infancy, but gaining momentum. Presently, less than 5 percent of global dysprosium is recovered from discarded devices, primarily because magnets are embedded within complex assemblies that are difficult to disassemble and purify.

However, new technologies and business models are shifting this paradigm. Start-ups like Ionic Technologies in Belfast and Geomega in Canada are developing chemical and electrochemical methods to recover dysprosium from used magnets with high efficiency. If scaled, these processes could provide hundreds of tonnes of secondary supply annually—critical not only for stabilizing the market but for reducing the environmental burden of virgin mining.

The market trends for dysprosium reflect its growing strategic value. As of April 2025, the price hovered around $203 per kilogram, but that figure had swung by as much as 45 percent over recent years. These fluctuations are not driven solely by economic demand, but also by political decisions. When China imposed new export restrictions in early 2025, global industries were jolted. EV manufacturers scrambled to secure supply, while governments re-evaluated their critical minerals strategies. This volatility underscores a deeper truth: in the age of electrification, materials like dysprosium are no longer mere commodities—they are geopolitical assets.

The supply-demand gap continues to widen, particularly as clean energy technologies scale. The annual growth rate of demand is estimated at 8 to 9 percent, outpacing the current pace of supply expansion. If this imbalance persists, dysprosium could become a bottleneck for global efforts to decarbonize and digitize economies. Recognizing this, countries like the United States, European Union members, Australia, and India are now pursuing robust strategies to build resilient supply chains. These include policy incentives for domestic mining, subsidies for refining plants, strategic stockpiling, and international partnerships aimed at resource cooperation and transparency.

For students and future professionals, this chapter provides more than facts—it offers a framework. Understanding how a single element like dysprosium flows through global systems reveals the interconnectedness of science, economy, and governance. It is a case study in the challenges of building a sustainable future when the materials we depend on are neither infinite nor evenly distributed. Dysprosium teaches us that the race for innovation begins long before a product is designed. It begins at the source—and the choices made there will shape the future for generations to come.

## Challenges & Opportunities in the Future of Dysprosium

The future of Dysprosium (Dy), a critical rare earth element used in high-performance magnets, is shaped by a complex interplay of environmental concerns, technological frontiers, and strategic innovation. As global demand rises for electric vehicles, wind turbines, and military-grade electronics, nations and industries must navigate the dual forces of opportunity and challenge.

Mining Dysprosium presents substantial environmental risks that cannot be overlooked. Extracting Dy typically involves large-scale open-pit mining operations, which often generate radioactive byproducts, particularly thorium and uranium traces found within the ore. These substances demand careful handling and long-term containment strategies. Additionally, the mining process results in acidic wastewater, which, if mismanaged, threatens surrounding ecosystems. The economic burden of safely managing these byproducts, alongside the increasingly strict environmental regulations in regions outside China, has significantly elevated operational costs. Moreover, companies face permitting delays and regulatory hurdles that can extend project timelines and diminish investor confidence. While China remains dominant in the rare earth supply chain, other countries striving for independence face these costly and time-consuming barriers.In response to these challenges, the concept of circularity and sustainability has emerged as a promising frontier. One of the most discussed approaches is the recycling of rare earth magnets, particularly those found in consumer electronics, electric motors, and industrial equipment. Recycling offers the potential to reduce dependency on virgin mining, alleviate environmental pressures, and create a stable domestic supply chain. However, scaling up recycling to meaningful levels poses its own set of hurdles. The collection and sorting of end-of-life products require robust infrastructure and consumer participation, both of which are lacking in most regions. Furthermore, the technical challenge of separating Dysprosium from other elements without significant material loss is still under active research. The economics of rare earth recycling remain marginal at best, and without major policy support or breakthroughs in extraction efficiency, it may struggle to meet ambitious legislative targets—such as the European Union's goal to recover 15% of rare earth elements through recycling by 2030. Still, intense R&D efforts continue globally, driven by the urgency of building a resilient and environmentally sustainable supply chain.

Simultaneously, scientific exploration into rare-earth-free alternatives is advancing at a steady pace. Researchers are experimenting with transition metal alloys, nanocomposites, and innovative magnetic materials that could replace or reduce dependence on Dysprosium. Yet, one critical hurdle persists: thermal performance.

Dysprosium's role in magnets is to enhance resistance to demagnetization at high temperatures—an essential trait for motors operating under extreme conditions, such as in electric vehicles or jet engines.

Current alternatives, while promising in controlled environments, often fail to match Dy-containing magnets when subjected to industrial heat and stress. This limitation reinforces Dysprosium's strategic importance, especially for defense and aerospace applications where reliability under duress is non-negotiable.

These three interlocking themes—environmental stewardship, recycling innovation, and substitute development—highlight a dynamic tension in the global pursuit of sustainable technology. For young minds aspiring to work at the crossroads of science, sustainability, and strategy, Dysprosium offers a case study rich in complexity and consequence. It is a symbol of both the costs and the promise of modern innovation, and those who can help solve its challenges will shape the material future of the planet.

**End of the Chapter**

# Chapter–3. The Green Element: Praseodymium's Role in Tomorrow's Technology

*Often overlooked yet essential to green innovation, praseodymium is quietly shaping the technologies that will define tomorrow. This chapter reveals how this rare earth element strengthens magnets, powers clean energy systems, and enables advanced optics and alloys. By understanding praseodymium's role, readers gain insight into the material foundations of sustainability, electrification, and the future of intelligent technology.*

Praseodymium, with atomic number 59, belongs to the group of light rare earth elements and is quietly revolutionizing many of the most advanced technologies in the world today. Though not commonly known outside scientific or industrial circles, its influence runs deep—particularly in the age of electrification, miniaturization, and intelligent machines. Its soft, silvery appearance belies the immense magnetic and catalytic properties it holds within its atomic structure. When combined with its rare earth sibling neodymium, it forms the powerful NdPr oxide, a cornerstone material in the production of NdFeB (neodymium-iron-boron) permanent magnets. These magnets are not merely components; they are the core enablers behind countless high-performance devices—ranging from electric vehicle motors to wind turbine generators—where efficiency, compactness, and durability are mission critical.

## What is Praseodymium & Why it Matters

Beyond magnets, Praseodymium is also celebrated for its versatility across scientific and industrial domains. Its optical properties make it a valuable tool in the design of specialty glasses and laser technologies.

In particular, it is used to dope certain crystals, enhancing their performance in fiber lasers and diode-pumped solid-state systems. Praseodymium's unique electronic structure allows it to emit a bright, stable light when excited—making it ideal in advanced optical communication systems and high-precision laser equipment. Furthermore, in the field of pigments, Praseodymium compounds produce brilliant yellow-green hues that are both vibrant and chemically stable. These are used to color ceramics and glass, often in applications requiring resistance to high temperatures and environmental degradation.

As a catalyst, Praseodymium proves its mettle in chemical reactions that demand durability and control under extreme conditions. It is particularly valued in refining processes and in the production of certain polymers and synthetic fuels, where it helps regulate molecular transformation with precision. These catalytic roles, though often overlooked by the public, are essential for energy efficiency and environmental sustainability in modern industry.

Praseodymium also plays a vital role in metallurgy, especially when alloyed with magnesium to create high-strength, lightweight materials. These alloys are invaluable in aerospace and automotive engineering, where reducing weight without compromising strength can dramatically improve fuel efficiency and performance. In turbine blades, engine components, and structural aerospace parts, Praseodymium contributes to creating materials that can endure intense heat and mechanical stress while remaining light and adaptable.

Its wide-ranging applications reveal the reason behind its growing strategic significance. In an age where the world is shifting toward renewable energy, electric mobility, and intelligent systems, the materials that power these transitions become vital to national economies and technological leadership. Praseodymium, though modest in visibility, is one such strategic material. Its properties—magnetic, optical, catalytic, and structural—place it at the heart of the innovation ecosystem, giving it a unique status among the rare earth elements.

For young learners and aspiring scientists, Praseodymium offers a powerful example of how subtle atomic variations can yield revolutionary outcomes. Understanding such elements is more than an academic exercise; it is a gateway into the future of sustainable technologies, clean energy, and smarter design. As the world accelerates toward digital transformation and climate resilience, the mastery of elements like Praseodymium will separate passive consumers from active inventors of tomorrow.

## Praseodymium in Action: From Earth's Crust to Global Industries

In the intricate web of modern technologies, few elements exert as broad and profound an influence as Praseodymium. Though it lies buried deep within the Earth's crust, its reach extends into the skies through aerospace innovations, across electric highways through EVs, and even into the precision of photonics and lasers. Understanding the core applications of Praseodymium is essential for grasping how the modern world is powered, shaped, and continuously reinvented.

The single most impactful use of Praseodymium today lies in the manufacture of high-performance permanent magnets, particularly for electric vehicles and wind turbines. Combined with neodymium, Praseodymium significantly enhances the strength, thermal stability, and reliability of NdFeB magnets—the strongest commercial magnets available. In electric motors, these magnets ensure efficient torque and speed control while minimizing energy loss. In wind turbines, they enable more compact and durable generators, allowing energy to be harvested even in remote or harsh environments. As global demand for clean energy and electric mobility accelerates, this application alone represents the largest and fastest-growing market segment for Praseodymium.

Another critical field lies in the realm of advanced alloys. When alloyed with magnesium, Praseodymium contributes to the development of lightweight, high-strength materials that withstand extreme temperatures and mechanical stress. These alloys are vital in aerospace engineering, where every gram of weight saved can lead to substantial improvements in fuel efficiency and maneuverability. Praseodymium-infused materials are now being used in aircraft components, turbine blades, and space-grade engineering systems—making them indispensable in the race for energy-efficient flight and space exploration. In optics and laser technology, Praseodymium again demonstrates its extraordinary versatility. It is a favored dopant in fiber amplifiers and solid-state lasers, where it enhances beam precision and stability. These applications are not limited to laboratories; they extend to telecommunications, medical equipment, and defense systems. Praseodymium-doped crystals are critical in generating specific light frequencies, supporting advanced imaging, laser surgery, and industrial cutting operations with pinpoint accuracy.

Even in everyday materials like glass and ceramics, Praseodymium leaves its mark—literally. Its compounds produce vivid yellow-green pigments used in specialty lenses and protective eyewear, such as welding goggles. These applications, while seemingly modest, are vital in ensuring safety and visual clarity in high-risk environments. The element's coloration properties are also prized in high-end art glass, optical filters, and temperature-resistant tiles.

While its applications continue to expand, so too does the global competition for its production. At the center of the supply landscape is China, which currently dominates the mining, separation, and refining of Praseodymium. Through decades of investment and infrastructure development, China has secured control over a vast portion of the world's rare earth output. This dominance allows it to influence pricing, availability, and strategic reserves—placing other nations in a position of dependency.

However, a notable shift is underway. The United States, with its vast reserves at Mountain Pass in California, is actively working to reestablish its rare earth industry. MP Materials, a key player in this movement, is expanding beyond raw extraction into refining and magnet manufacturing—efforts backed by significant government support and defense contracts. New facilities in Texas and Fort Worth signal a broader strategy to build a domestic rare earth supply chain, reducing reliance on foreign imports for critical technologies.

Australia has also emerged as a powerful contributor, with Lynas Rare Earths operating one of the richest deposits at Mt Weld. Its refining and separation operations, though partially conducted overseas, are increasingly shifting toward self-sufficiency. In parallel, the United Kingdom is investing in facilities like Pensana's Salt End, designed to process NdPr oxide and support European magnet production. These developments reflect a broader global realization: the path to clean energy and advanced industry must include stable, diversified sources of critical elements like Praseodymium. For students, scientists, and policymakers alike, the story of Praseodymium is a lesson in how raw materials become national assets and global leverage points. It reminds us that the future of green technology, aerospace innovation, and digital infrastructure often begins with the strategic extraction and application of elements most people have never heard of. And for those who seek to build, lead, and innovate in the 21st century, knowledge of such elements is no longer optional—it is essential.

## Praseodymium Economics: Markets, Momentum, and the Materials Race

The rise of Praseodymium from an obscure rare earth element to a linchpin of modern industry is not just a technological story—it's also a tale of economics, policy, and strategic risk. As the world accelerates toward clean energy and smart technology, the demand for high-performance magnets and specialty materials is rewriting the rules of global resource markets. Praseodymium, often in combination with

Neodymium, is at the very center of this shift, its value increasing not just in dollars, but in geopolitical weight.

By 2025, the global market for Praseodymium is projected to reach approximately $1.5 billion, largely driven by the booming electric vehicle (EV) and wind energy sectors. These industries, hungry for compact and efficient motors and generators, rely heavily on NdFeB (Neodymium-Iron-Boron) permanent magnets, where Praseodymium plays a key strengthening role. From vehicle propulsion systems to offshore wind turbines, these magnets provide the performance edge that modern systems demand. This application alone is fueling one of the most significant rare earth market expansions in recent history.

The trend is not temporary. According to long-term projections, demand for NdFeB magnets—and by extension, Praseodymium—could increase by a factor of four to seven by 2040. This growth isn't speculative; it reflects committed industrial strategies, government policy roadmaps, and a cultural pivot toward electrification. Even in 2024–25, steady market expansion was already observable, signaling not just a recovery from prior lows but a sustained upward trajectory.

Corresponding with rising demand is a notable shift in pricing trends. For several years, PrNd alloy prices had languished due to oversupply and soft market conditions. But that era appears to be over. With China tightening its control over supply chains and global consumption on the rise, prices are rebounding sharply. Analysts anticipate that elevated pricing will continue through late 2025 and possibly beyond, driven by both strategic stockpiling and industrial scaling. For industries reliant on Praseodymium-based magnets, these price movements present both cost challenges and incentives to explore alternatives.

However, price is just one dimension of a more complex strategic landscape. Perhaps the most significant risk facing the

Praseodymium market is supply concentration. Today, China controls the vast majority of global mining, separation, and refining capacity for rare earth elements. This dominance gives it tremendous influence over availability, pricing, and international policy leverage. For nations outside China, especially those prioritizing technological independence, this creates a geopolitical vulnerability.

In response, a number of strategic initiatives are underway. The United States has ramped up its support for domestic rare earth projects, most notably through Pentagon-backed investments in MP Materials. These efforts aim not just to extract raw materials, but to establish a full magnet supply chain—from mine to motor—within U.S. borders. Australia, Canada, and select European nations are also stepping up exploration and processing efforts to reduce dependence on Chinese supply lines.

Part of the diversification strategy also includes recycling. Rare earth magnet recycling is an emerging field with significant potential. If scaled effectively, recycling could supply up to 15% of global Pr and Nd demand by 2030.

The environmental and economic benefits of this approach are significant—it reduces the need for new mining, minimizes waste, and can stabilize prices over time. Yet, technical and logistical barriers remain. Efficient recovery systems must be designed, and consistent supply chains of end-of-life magnets must be secured.

Environmental considerations are another growing pressure point. Mining and refining rare earth elements, including Praseodymium, involve significant ecological challenges—radioactive waste, water pollution, and energy consumption chief among them. In Western countries, where environmental regulations are stringent and community engagement is critical, gaining permits requires demonstrable commitment to sustainable practices. Companies that fail to meet these expectations risk not only delays but public opposition. Meanwhile, researchers are actively pursuing substitute materials that could replicate the magnetic properties of Praseodymium-based alloys. Although these innovations show promise in certain applications, they have yet to reach the performance and temperature resilience that Praseodymium delivers—especially in high-stress industrial uses like EV motors and defense systems. As of now, no viable replacement matches the versatility and efficiency of Praseodymium in critical technologies.

Ultimately, the market story of Praseodymium is one of both high stakes and high opportunity. For youth exploring the intersection of science, economics, and environmental responsibility, this element offers a powerful example of how deeply material science influences global systems. The rare earth race is not just about mining minerals—it's about securing the future of energy, transportation, defense, and digital infrastructure. In this race, knowledge is the most valuable resource of all.

**End of the Chapter**

# Chapter-4.
# Terbium:
# Lighting the Future and Powering Precision

*Terbium is a rare element with an outsized impact—quietly illuminating our screens, strengthening clean energy systems, and enabling precision technologies vital to modern life. This chapter explores how Terbium's unique optical and magnetic properties make it indispensable to lighting, renewable energy, and defense, revealing why this little-known element sits at a critical crossroads of innovation, strategy, and sustainability.*

In the world of advanced materials, few elements bridge the gap between vibrant illumination and strategic defense quite like Terbium. With atomic number 65 and symbol Tb, this rare-earth metal belongs to the lanthanide series and exhibits a fascinating combination of magnetic, optical, and structural properties. Though it never appears in its pure form in nature, Terbium is extracted from complex mineral ores like monazite and xenotime—minerals that house several rare earth elements in a tightly interlocked geochemical matrix.

## What Terbium is & Why it's Crucial

Terbium's significance stems from two primary qualities: its remarkable luminescence and its ability to enhance magnetic behavior under thermal stress. These traits have propelled it into technologies at the very core of modern life—from the screens we stare at to the turbines that power our cities. Despite its low profile in mainstream discussions, Terbium is one of the most essential building blocks of the digital and clean-energy revolutions.

A major portion of Terbium's global use—roughly three-quarters—goes into the production of green phosphors, which are indispensable for flat-panel displays, LED lighting, and compact fluorescent lamps. Terbium oxide enables these phosphors to produce vivid, stable, and energy-efficient green light. It's what gives digital screens their balance and clarity, ensuring color accuracy across televisions, smartphones, and computer monitors. As consumers and industries push for better visuals and reduced energy consumption, this application continues to dominate the Terbium market.

Yet, Terbium's role does not end at lighting. It plays an equally vital part in the magnetic systems that drive electric vehicles and wind turbines. When added to NdFeB (neodymium-iron-boron) magnets, Terbium improves coercivity—the ability of a magnet to retain its magnetization under heat and external forces. This is critical for technologies that must operate in high-temperature environments, such as wind turbine generators or electric vehicle motors that endure intense mechanical and thermal stress. For instance, a single 3-megawatt wind turbine can require up to two kilograms of Terbium, while electric motors in vehicles may need up to one kilogram per unit. These are not incidental quantities—they reflect how deeply embedded Terbium has become in the machinery of the green energy transition.

Beyond energy and electronics, Terbium also plays a strategic role in defense and aerospace. It is a key ingredient in magnetostrictive alloys such as Terfenol-D, which change shape in response to magnetic fields. These unique materials are used in underwater sonar systems, precision-guided weapons, and aerospace sensors, where exactitude and reliability are non-negotiable. Because of its utility in these high-stakes applications, the U.S. Department of Defense classifies Terbium as a Tier-1 strategic material—essential to national security and military innovation.

What makes Terbium especially fascinating for young scientific minds is its emerging potential in frontier technologies. In the medical field, it is being explored as a contrast agent in MRI imaging due to its favorable magnetic resonance properties. In quantum computing, Terbium-doped materials are showing promise in sensing and isolating quantum states—an essential function for building error-resistant quantum bits. In biomedical imaging, Terbium-based probes are being researched for their ability to fluoresce with high specificity, offering new possibilities in diagnostics and targeted treatment.

From lighting up the world's screens to operating deep-sea sensors, from enabling efficient clean energy to shaping the future of quantum medicine, Terbium is a master of adaptation and precision. Its story teaches us that true innovation often depends not on the most famous elements, but on the quiet contributors whose unique capabilities make the impossible practical. For students and young explorers of science, Terbium stands as a reminder that every element has a role—and some, like this silvery-white metal, are shaping the course of tomorrow in more ways than one.

Terbium's Global Journey: From Mines to Markets Understanding where Terbium comes from, how it is processed, and what affects its pricing reveals the deeper story of a material that is both geologically rare and geopolitically significant. Unlike many common metals, Terbium belongs to the group of heavy rare earth elements (HREEs), which are more difficult to extract and purify due to their scattered presence in ores and complex separation processes. Its value is not only defined by its physical properties but also by the limited number of regions capable of producing it at scale.

At the center of the Terbium supply chain is China, which controls approximately 90–95% of global heavy rare earth processing capacity. This dominance is not accidental—it is the result of decades of investment in refining infrastructure, resource access, and export strategy. China's refining hubs are highly specialized, capable of separating trace amounts of Terbium from ore concentrates that often contain many rare earth elements in tangled molecular structures. As a result, most of the world's Terbium passes through Chinese refineries before it reaches manufacturers across the globe.

However, much of the raw material that feeds into China's supply chain originates beyond its borders. Myanmar, particularly the mineral-rich Kachin State, has become a crucial supplier of heavy rare earth ores, contributing between 14% and 57% of the feedstock used by Chinese processors. This dependency on Myanmar introduces significant volatility into the market. Political instability, environmental degradation, and conflict in mining regions frequently disrupt extraction and transport, causing uncertainty in the global Terbium flow. These disruptions reverberate across industries that rely on steady, reliable access to critical materials.

In contrast, countries like Australia are working diligently to emerge as stable and sustainable producers of heavy rare earths, including Terbium. Projects such as Browns Range, developed by Northern Minerals, and Iluka Resources' Eneabba refinery are designed to produce high-purity Terbium oxide from Australian soil. These efforts are strongly backed by public funding—amounting to nearly a billion U.S. dollars—reflecting a strategic commitment to reducing dependence on volatile or monopolized supply chains. Australia's entry into the Terbium market marks a broader push by Western nations to localize critical mineral production and establish secure, ethical alternatives.

Meanwhile, recycling is being explored as a complementary source of Terbium. Although the volume currently recovered from electronic waste and discarded magnets remains under 3% of global supply, it represents a critical avenue for future resilience. Terbium recovery from used devices is technically challenging—often requiring intricate dismantling and separation steps—but the environmental and economic advantages are undeniable. As demand rises and sustainability becomes a central concern, innovations in recycling could help balance supply constraints and reduce reliance on raw mining.

Market trends offer additional insight into Terbium's strategic importance. Annual production of Terbium oxide currently stands at around 400 to 450 metric tons. However, demand is projected to grow sharply, potentially reaching 700 tons by 2028. This widening gap between supply and demand exerts direct pressure on prices. Recently, Terbium prices have experienced significant spikes, ranging between $1,800 and $2,400 per kilogram. These increases are largely driven by two key factors: export controls imposed by China and intermittent supply disruptions in Myanmar. Such conditions create an environment where market prices can change rapidly, affecting downstream industries from electronics to clean energy systems.

For students, innovators, and future policymakers, the Terbium supply chain offers a revealing case study in how natural resources shape global dynamics. It shows how the journey of a single element—from obscure mineral deposits to high-stakes markets—can influence energy policy, technological advancement, and international cooperation. The choices we make today about how we mine, recycle, and distribute Terbium will have lasting consequences, not just for industrial efficiency but for sustainability and global stability. In this way, Terbium isn't just a material of the future—it is a mirror reflecting the priorities and challenges of our time.

## Terbium Under Pressure: Global Risks and Strategic Crossroads

The story of Terbium is not just one of innovation and opportunity—it is also a tale marked by fragility, ethical tension, and global power dynamics. As demand for this heavy rare earth element continues to surge across the clean energy, electronics, and defense sectors, the challenges associated with securing a stable, ethical, and sustainable supply chain have come sharply into focus. For nations, industries, and individuals shaping the future, understanding these risks is essential.

In early 2025, the world witnessed a dramatic reminder of how geopolitics can shape the material foundation of technology. In Myanmar's Kachin State, which supplies a significant portion of the raw feedstock for China's heavy rare earth refining industry, a sudden rise in militia activity brought mining operations to a near standstill. Given that Kachin contributes between one-third and one-half of global heavy-REE inputs, the disruption was severe. In response, China—the primary processor of Terbium—issued export ultimatums that rippled through global markets, threatening to restrict access for many Western economies.

This event exposed a fundamental vulnerability: too much of the world's critical mineral supply is concentrated in politically unstable or monopolized regions. To counter this, Australia and its allies have taken assertive steps. Projects led by companies like Iluka Resources, Lynas Rare Earths, and Northern Minerals are being accelerated, supported by government initiatives and financial backing. Some nations are also establishing strategic stockpiles of Terbium and other rare earth elements, a move once reserved for oil or grain reserves, now repurposed for the age of green technology.

However, geopolitical instability is not the only challenge facing the Terbium supply chain. Equally troubling are the environmental and ethical consequences associated with its extraction. In Myanmar, the dominant method for heavy rare earth mining is in-situ leaching—a process that injects chemical solutions into the earth to dissolve rare earth elements for extraction. While efficient, this method leaves a toxic legacy. Chemicals like oxalic acid, used in the process, often seep into surrounding soils and waterways, posing risks to human health, agriculture, and biodiversity. Entire communities have reported water contamination and ecosystem damage, fueling resistance to mining expansion and international criticism.

This raises urgent ethical questions for the industries and consumers that benefit from Terbium. As awareness of these impacts spreads, companies are facing intensifying pressure from governments, watchdog organizations, and environmentally conscious consumers to adopt responsible sourcing practices. Transparency in supply chains, environmental compliance, and human rights oversight are no longer optional—they are becoming defining factors of market legitimacy and social license to operate.

For young leaders and future scientists, this is a defining moment. Terbium is a window into the complex tradeoffs behind every piece of advanced technology: the tension between innovation and sustainability, security and cooperation, progress and accountability. Navigating this terrain demands more than technical expertise—it requires ethical clarity, strategic vision, and a willingness to rethink how resources are valued and managed. As nations race to build the technologies of tomorrow, the deeper question becomes: who will build them responsibly? The answer will shape not only the future of Terbium, but the integrity of the systems it empowers.

**End of the Chapter**

# Chapter–5.
# Cerium and Lanthanum: The Quiet Workhorses of Modern Industry

*Cerium and Lanthanum rarely command attention, yet they quietly sustain the machinery of modern life. This chapter uncovers how these abundant rare earth elements underpin catalysis, clean fuels, precision optics, and energy storage. By exploring their industrial roles and global pathways, readers gain insight into the unseen materials that keep modern economies efficient, cleaner, and technologically resilient.*

Among the rare earth elements, Cerium and Lanthanum rarely capture the spotlight—but their significance is nothing short of foundational. As light rare earth elements, or LREEs, they occupy the beginning of the lanthanide series and dominate in both abundance and utility. Together, Cerium (Ce) and Lanthanum (La) account for nearly 70% of all separated rare earth oxides found in commercial use, a remarkable statistic that highlights their central role in shaping modern industrial processes.

## What they are & Why they Matter

Though these elements are more geologically abundant than many everyday metals, their real challenge lies not in scarcity, but in separation. They are typically locked together with other rare earths in complex mineral ores, such as bastnäsite and monazite. Extracting them in commercially pure forms requires intricate chemical refinement, sophisticated separation techniques, and a well-developed supply chain—making their availability more dependent on technology and infrastructure than geology.

Cerium and Lanthanum are not prized for their glamour, but for their extraordinary functionality across a wide range of industries. Nowhere is this more evident than in the field of catalysis, particularly in oil refining and automotive emission control. Cerium oxide plays a critical role in fluid catalytic cracking (FCC), a process that increases gasoline yield from crude oil by as much as 7 to 10 percent per barrel. This enhancement, though invisible to the end user, translates into enormous gains in fuel efficiency and industrial output. In automotive systems, both Cerium and Lanthanum are integral to the operation of catalytic converters, devices that significantly reduce harmful vehicle emissions. As global car production continues and emissions regulations tighten, the demand for these materials only intensifies.

Another area where Cerium excels is in the precision world of glass polishing. Cerium oxide, commonly known as ceria, is the gold standard for polishing high-value glass surfaces. Whether it's the display of your smartphone, the lens of a camera, or the screen of a flat-panel television, ceria is likely responsible for the final optical clarity and smooth finish. Its fine abrasiveness, combined with chemical activity, enables unparalleled polishing results, making it indispensable in electronics manufacturing and optical fabrication.

Lanthanum, while equally industrial in scope, distinguishes itself in specialty optics. Lanthanum oxide is a key ingredient in the production of high-refractive-index glass, which is used in camera lenses, microscopes, infrared sensors, and even astronomical telescopes. This glass doesn't just bend light more effectively—it does so while minimizing distortion and chromatic aberration, leading to sharper, clearer images in both consumer electronics and scientific instruments.

The energy storage sector also benefits from Lanthanum's properties. In particular, Lanthanum-based alloys form the negative electrode in nickel-metal hydride (NiMH) batteries. These batteries have powered millions of hybrid vehicles globally, including early models of the Toyota Prius, and continue to find relevance in tools, electric bicycles, and backup power systems. While lithium-ion technologies dominate current headlines, NiMH batteries remain crucial in markets requiring safety, cost-effectiveness, and longevity.

Cerium and Lanthanum extend their utility further into emerging and specialized domains. Cerium compounds are used in electrodes for gas tungsten arc welding, offering stability under high-temperature arcs.

They also help in decolorizing glass and filtering ultraviolet light, protecting both materials and human vision from UV exposure. Lanthanum, on the other hand, finds applications in radiation detection. Its compounds are used in scintillators—materials that detect gamma rays and neutrons—playing vital roles in nuclear imaging, border security, and research instrumentation.

Additionally, Lanthanum contributes to high-tech ceramics and advanced lighting systems.

For students, engineers, and future innovators, Cerium and Lanthanum serve as a lesson in the power of utility. They are not flashy elements, but they are irreplaceable. They teach us that scientific importance is not always about rarity—it is about the depth and diversity of impact.

These quiet workhorses of the periodic table enable progress across energy, environment, electronics, and optics, forming the hidden foundation of many technologies we now take for granted. To understand them is to appreciate the materials science that supports the infrastructure of our modern world.

## The Global Path of Cerium and Lanthanum: From Mine to Market

To understand the true value of Cerium and Lanthanum, one must go beyond their uses and explore where they come from and how they move through the global economy. These light rare earth elements may be geologically abundant, but the challenge lies in their extraction, refinement, and consistent supply. Their production landscape is not just a story of geology—it's one of geopolitical strategy, technological leadership, and global trade dynamics.

China stands at the center of this complex map. For Cerium and Lanthanum, China accounts for roughly 60% of the world's production, a figure that mirrors its dominance in overall rare-earth mining. But the influence doesn't stop at extraction. More significantly, China controls over 85–90% of the global refining capacity—a stage that turns raw mineral concentrates into the purified compounds needed for industrial use. This refinement supremacy gives China significant leverage over global supply chains and pricing, making it a pivotal player in the rare-earth ecosystem. While China may lead, other nations contribute strategically. The United States, for instance, produces about 10% of the global rare earth supply, largely through the Mountain Pass mine in California. Though rich in light rare earths, Mountain Pass historically relied on China for separation and refining. However, new efforts are underway to create a full domestic supply chain—from mining to finished products—highlighting a renewed commitment to material independence.

Australia, India, and Russia also play essential roles in the global rare-earth economy. These countries are especially active in the extraction of mixed rare-earth ores, which include Cerium and Lanthanum alongside several other lanthanides. India, in particular, holds a unique position. Through its state-run Indian Rare Earths Limited (IREL), it processes monazite-rich beach sands to extract Ce and La compounds. These materials are then exported to technology hubs across the world, supporting industries that rely on clean energy, electronics, optics, and advanced manufacturing.

All of this activity is driven by an expanding market. In 2025, the global rare-earth element market was valued at approximately $6 billion—a significant figure for materials once considered too obscure to matter commercially. Looking ahead, projections suggest this market could grow to between $9 and $11 billion by the early 2030s, fueled by demand across renewable energy, electric vehicles, digital displays, and industrial automation.

Within this growing market, light rare earth elements like Cerium and Lanthanum continue to dominate in terms of volume. While heavy rare earths are seeing faster growth due to their importance in magnetic technologies, the sheer scale of applications for Cerium and Lanthanum keeps them at the top in terms of tonnage used. From automotive catalysts to glass polishing, battery components to specialty optics, these elements form the industrial backbone of multiple sectors.

One revealing indicator of Cerium's market impact is the niche but significant segment of ceria polishing powders. In 2024 alone, this product category generated approximately $1.1 billion in global revenue. With continued growth in electronics and optics manufacturing, forecasts suggest this number may climb to nearly $1.8 billion by 2033—a clear sign that even seemingly small applications can carry massive economic weight when scaled across global industries.

For the curious minds exploring the material world, Cerium and Lanthanum offer a compelling case study. They show how elements hidden in mineral sands can shape billion-dollar markets and influence international strategy. From refining control to rising demand, from the deserts of China to the beaches of India, the journey of these elements is deeply interwoven with the technological and economic story of the modern age. Understanding their production and market dynamics reveals just how much the future of innovation depends on the smart and ethical handling of Earth's most essential building blocks.

## Navigating Challenges and Charting the Future of Cerium and Lanthanum

Despite their wide-ranging uses and economic significance, Cerium and Lanthanum face a series of complex challenges that demand careful strategy, innovation, and global cooperation. The very processes that make these elements accessible—mining, extraction, and chemical refinement—carry with them substantial environmental costs.

Waste from these operations often contains toxic substances and, in some cases, low-level radioactive material, depending on the host ore. Without rigorous safeguards, these byproducts can contaminate soil and water systems, endangering local ecosystems and communities. As a result, regulatory pressures are rising, and companies are increasingly held accountable not just for production output but for sustainability and ethical stewardship.

Geopolitically, another layer of risk lies in the overconcentration of supply. With China maintaining dominance over both raw production and refining capacity, the global supply chain for Cerium and Lanthanum remains vulnerable to trade tensions, policy shifts, or disruptions in regional stability.

Recognizing this imbalance, several countries—including the United States, members of the European Union, India, and Australia—are investing heavily in rare-earth independence. These initiatives include funding new mining operations, expanding local refining facilities, and enhancing research into alternative sources and methods. While full diversification may take years, the direction is clear: resilience must replace reliance.

One of the more promising solutions emerging in this context is recycling. Unlike rare earth elements used exclusively in magnets—which are difficult to extract from end-of-life products—Cerium and Lanthanum are widely dispersed in materials like glass, automotive catalysts, and battery alloys. This opens doors to recycling pathways that are both technically feasible and environmentally responsible.

As electronic waste continues to rise globally, recovery of these materials offers an increasingly viable and valuable contribution to the circular economy. It also aligns with growing consumer and industrial demand for greener supply chains and reduced material footprints.

Looking forward, the future for Cerium and Lanthanum remains one of steady growth and strategic importance. As the world accelerates toward clean energy, sustainable infrastructure, and digital connectivity, demand for these elements is projected to grow at an annual rate of approximately 4 to 7 percent through the early 2030s. This growth will not be evenly distributed; it will be shaped by industrial policy, regional capability, and technological leadership.

In North America, government-backed initiatives—especially those led by the U.S. Department of Energy—are scaling up domestic rare earth processing capacity, while universities and research institutions explore more efficient and less polluting refining technologies. In the Asia-Pacific region, production continues at pace, supported by established infrastructure and growing local demand. Europe, meanwhile, is focusing on building traceable, ethical, and low-emission rare earth supply chains to align with its broader climate goals.

Technological trends will further redefine how Cerium and Lanthanum are sourced and used. Artificial intelligence is being deployed to optimize mining operations, helping identify richer ore deposits and reduce waste. Cleaner chemical processes are being introduced to lower the environmental burden of refining. Recycling techniques are being refined to recover polishing powders from manufacturing waste and catalysts from spent automotive components—signaling a shift from linear to circular material lifecycles.

For today's youth—the scientists, engineers, and policymakers of tomorrow—Cerium and Lanthanum offer a blueprint for navigating complexity with creativity. They illustrate how deeply materials are woven into the challenges of our time and how solutions will require not just technological prowess, but ethical commitment and global vision. In understanding these elements, students do more than study chemistry—they learn how the material world must be managed to serve both progress and the planet.

## End of the Chapter

# Chapter–6.
# Yttrium:
# The Unsung Architect of Advanced Technology

*Often hidden behind brighter names, Yttrium and Scandium quietly shape the frontiers of modern technology. This chapter reveals how these unsung elements enable lasers, superconductors, advanced alloys, clean energy systems, and medical breakthroughs. By exploring their roles and strategic importance, readers gain insight into the subtle materials that make high-performance, sustainable innovation possible.*

Among the rare earth elements that quietly power the technologies of the future, Yttrium holds a distinctive place. With atomic number 39, Yttrium is often grouped with the lanthanides due to its similar properties and geological behavior. Though not a lanthanide by classification, it is always found in the same mineral families, like xenotime and monazite, and never occurs in pure metallic form in nature. Its journey from a buried oxide in obscure mineral veins to a core material in lasers, LEDs, and cancer therapies is a testament to how modern science transforms the hidden into the indispensable.

## Why Demand is High

One of Yttrium's most impactful contributions lies in the realm of light—specifically, how it helps create and control it. Yttrium aluminium garnet (YAG), often doped with cerium or neodymium, is foundational in white LED production and high-brightness display technology. These phosphors are responsible for transforming blue LED light into a broad-spectrum white light that mimics natural daylight, making screens brighter, colors truer, and energy usage more efficient. Yttrium iron garnet (YIG), another derivative, is prized for its role in acoustic and microwave signal processing, enabling precise filters in telecommunications and scientific instruments.

Yttrium also plays a critical role in laser technology. YAG lasers, frequently doped with neodymium (Nd:YAG), are used in everything from industrial cutting and welding to medical surgery and cosmetic procedures. These lasers offer precision, power, and durability, helping industries push the limits of manufacturing and medicine. In another domain altogether—superconductivity—Yttrium emerges again. In materials like $YBa_2Cu_3O_7$ (commonly abbreviated as YBCO), Yttrium enables the phenomenon of high-temperature superconductivity. These materials conduct electricity with zero resistance under cryogenic conditions and are the subject of cutting-edge research into quantum computing, magnetic levitation, and ultra-efficient energy systems.

The medical world, too, finds great value in Yttrium. A radioisotope of this element, Yttrium-90, has revolutionized targeted cancer therapies. In procedures like radioembolization, microscopic Yttrium-90 particles are injected directly into tumors—particularly in liver cancer—delivering high doses of radiation to diseased tissue while sparing surrounding healthy organs. It's a powerful example of how nuclear science and biology can intersect to create life-saving innovation.

Even in the realm of metallurgy and ceramics, Yttrium continues to assert its quiet strength. It's used to stabilize high-temperature ceramic materials and coatings, which are essential in jet engines, gas turbines, and heat-resistant filters. In steel production, trace amounts of Yttrium improve grain refinement and oxidation resistance, enhancing durability and performance under stress. These uses, though less visible to the public, are foundational to industrial infrastructure and high-tech engineering.

Geopolitically, the production of Yttrium reflects many of the same patterns seen with other rare earths. China maintains a commanding position, responsible for roughly 60–70% of global rare earth mining and most of the world's Yttrium refining capacity.

Yet, other countries including the United States, Australia, Brazil, India, Russia, Vietnam, and Myanmar possess significant reserves and are working to expand their roles in both mining and processing. This effort is not just about resource extraction—it's about securing technological sovereignty in an increasingly competitive global economy.

From a market perspective, the story of Yttrium is one of steady and strategic growth. In 2024, the Yttrium metal market was valued at approximately $400 million. Projections suggest this could rise to around $580 million by 2030, driven by a compound annual growth rate of about 5%. What fuels this trajectory is not merely a growing population or consumption—it's a deepening dependence on digital infrastructure, artificial intelligence, medical innovation, and energy resilience. Yttrium, in its many forms, supports all of these.

For youth today—future engineers, scientists, medical researchers, and policymakers—Yttrium presents a compelling example of how one element can shape many industries.

It shows that behind the devices we hold, the lasers we rely on, and the treatments we hope for, are materials with extraordinary capabilities and complex supply chains. Learning about Yttrium is not just learning about an element; it is learning how the modern world is built, one invisible force at a time.

## Scandium: The Silent Catalyst of the Lightweight Revolution

Though often overlooked in the periodic table, Scandium—a trace element with atomic number 21—is fast emerging as a cornerstone of modern material science. Found only in minute concentrations—about 16 parts per million in the Earth's crust—Scandium rarely occurs in concentrated deposits. Instead, it is typically recovered as a secondary product from uranium processing or rare earth element (REE) mining. Yet, despite this elusive presence in nature, its impact across advanced engineering, clean energy, and experimental physics is growing rapidly.

The global production of scandium oxide remains relatively modest—between 15 to 20 metric tons per year. But these few tons are extraordinarily valuable, thanks to the element's unique ability to transform materials, especially aluminum. When combined with aluminum in small quantities, Scandium forms an alloy that is stronger, more weldable, and highly resistant to corrosion and fatigue. This Al-Sc alloy is particularly attractive to the aerospace industry, where reducing weight without sacrificing structural strength is crucial. From aircraft frames to unmanned aerial vehicles and space structures, these alloys are enabling a new generation of lighter, more efficient flying machines. The same properties make them a favorite in high-end sports equipment, such as premium bicycle frames, baseball bats, and even specialized automotive components.

Scandium's value, however, extends far beyond the skies. One of its most important emerging roles lies in solid oxide fuel cells (SOFCs), a technology at the forefront of clean energy conversion. These fuel cells generate electricity through electrochemical reactions, and when doped with $Sc_2O_3$ (scandium oxide), the ceramic materials used in their cores become significantly more conductive to oxygen ions. This improves efficiency, reduces operating temperatures, and extends the lifespan of the cells—making them more viable for powering buildings, remote facilities, and even next-generation electric vehicles. As global attention shifts toward decarbonization and sustainable energy infrastructure, the demand for Scandium in SOFCs is expected to accelerate sharply.

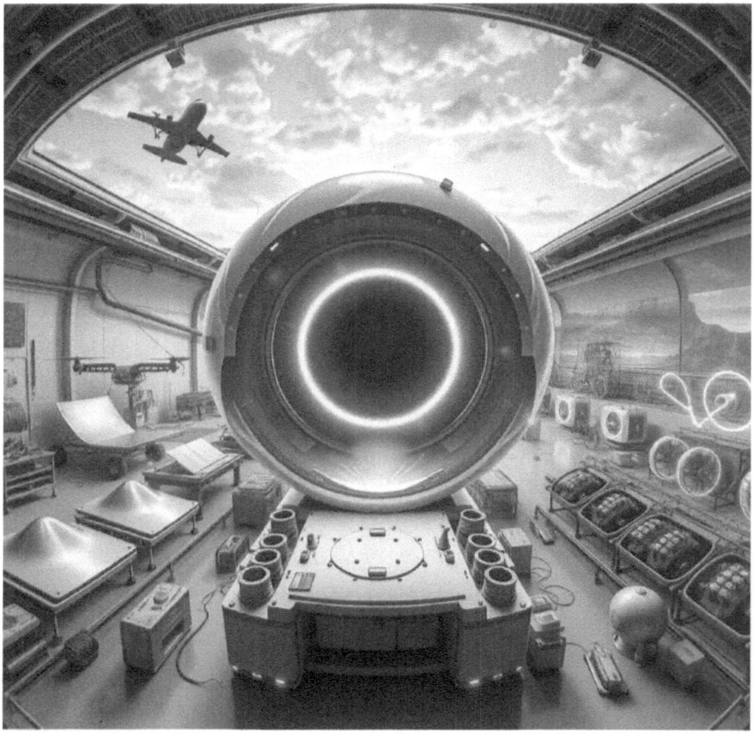

Another exciting application for Scandium is in additive manufacturing, more commonly known as 3D printing. Scandium-containing powders are now being used to produce high-performance aerospace and automotive parts. These powders offer not only excellent mechanical strength and thermal stability but also reduce the number of assembly steps required in traditional manufacturing. For complex or lightweight designs that must endure extreme environments, such as turbine blades or vehicle structural supports, Scandium-enhanced materials provide a powerful edge.

In the domain of scientific research, Scandium is also making intriguing appearances. Under high-pressure experimental conditions, pure Scandium has demonstrated superconducting behavior at temperatures as high as 36 Kelvin. While this remains in the early stages of study, it opens the door to potential breakthroughs in superconducting materials that could one day revolutionize electronics, transportation, and quantum computing.

The geographical footprint of Scandium production is concentrated, yet expanding. China is currently the largest producer, extracting it primarily as a byproduct from REE ore processing. Historically, Ukraine and Russia were among the first nations to commercialize Scandium production, with operations in regions like Zhovti Vody and the Kola Peninsula. More recently, attention has shifted to newer sources. The Philippines, for instance, is extracting Scandium from nickel-laterite tailings, while Australia is home to promising projects like the Clean TeQ Sunrise and Nyngan developments. These facilities, along with the Elk Creek project in the United States, are poised to significantly increase annual global output—potentially by up to 95 metric tons. Additional reserves are being explored in Madagascar, Norway, and other regions, pointing toward a future of more diversified supply.

All of these developments align with a robust market trajectory. In 2024, the global Scandium market was valued at roughly $750 million. Forecasts suggest this figure could rise to over $2.1 billion by 2032, reflecting a compound annual growth rate of nearly 14%. This is one of the fastest growth curves among rare and strategic metals, driven by advances in clean energy, materials engineering, lightweight design, and biomedical device manufacturing.

For young learners and future innovators, Scandium represents more than just a niche material—it is a gateway to understanding how small additions at the atomic level can reshape entire industries. It teaches us that technological revolutions are not always powered by abundant resources, but often by rare ones used wisely. In mastering the science and strategy behind Scandium, we glimpse the future of sustainable engineering and intelligent innovation—where performance meets precision, and efficiency meets elegance.

## Strategic Stakes and Future Pathways: Yttrium and Scandium in a Changing World

In the evolving landscape of advanced materials and critical technologies, few elements command as much strategic curiosity and potential as Yttrium and Scandium. They may be obscure to the average consumer, but within high-performance industries—spanning clean energy, digital infrastructure, aerospace, and medicine—they serve as foundational elements. However, behind their remarkable utility lie equally complex challenges, most notably related to supply, scalability, and sustainability.

The most immediate strategic concern for both Yttrium and Scandium is supply concentration. China continues to dominate the extraction and processing of Yttrium, controlling a substantial portion of the global output—nearly 70 percent. Scandium presents an even more constrained picture. It is not mined independently in large volumes, but rather collected as a byproduct from other mining operations, especially those focused on rare earth elements or uranium. This dependence on secondary recovery means that Scandium's availability is not based solely on its own demand but is linked to the economics of entirely different commodities. When production of those host minerals slows or shifts, Scandium supply contracts as a consequence. Adding to this fragility is the issue of cost. Scandium remains one of the most expensive elements per kilogram, with limited global output driving prices into the high hundreds or even thousands of dollars per kilo, depending on purity and form. Yttrium, while comparatively less expensive, is experiencing rising demand pressures that are beginning to affect its pricing trajectory. The result is a global market where demand is growing faster than supply can comfortably accommodate—especially as industries seek greener, lighter, and more efficient alternatives to traditional materials.

Nevertheless, the promise these elements hold is immense, particularly when one looks at where they are headed. For Yttrium, the continued advancement of superconductors—particularly YBCO compounds—offers revolutionary potential in energy transmission, quantum computing, and levitating transportation systems. Meanwhile, Scandium's integration into aluminum alloys is transforming the structural landscape of aviation, aerospace, and next-generation vehicles. These alloys are not only lighter and stronger but also more corrosion-resistant and weld-friendly, making them ideal for both performance and sustainability. In the energy domain, solid oxide fuel cells enriched with Scandium are shaping the frontier of decentralized, low-emission power solutions.

Recognizing both the strategic importance and the risks of over-dependence, countries across the world are moving swiftly to diversify their supply chains. Australia has emerged as a hub of new Scandium projects, while the Philippines explores tailings from nickel-laterite operations as an unconventional but viable source. In North America, production capabilities in the United States and Canada are being revitalized, and Ukraine continues to explore its legacy deposits. These global efforts signal a broad and urgent drive toward resilience, aiming to ensure that future supply can meet technological ambition without geopolitical bottlenecks. However, strategic expansion must be paired with environmental responsibility. The extraction and refinement of both Yttrium and Scandium involve complex chemical processes that can generate toxic waste, including heavy metals and acids. This environmental burden is now drawing significant attention, not only from regulators but also from investors and the public. In response, research into low-impact extraction methods, closed-loop recycling systems, and green chemistry solutions is accelerating. The recycling of materials from LED waste, spent catalysts, ceramics, and even old fuel cell stacks represents an emerging opportunity to recapture value and reduce ecological pressure—especially in nations without native ore deposits.

Taken together, the story of Yttrium and Scandium is one of high stakes and high promise. Both elements sit at the intersection of scientific discovery and strategic necessity, and both are reshaping how industries approach innovation, performance, and sustainability. Whether used in the silent heart of a superconducting magnet or the structural skeleton of a space-bound vehicle, they carry with them the burden and the potential of progress.

For youth seeking to understand the material forces behind the technologies they use—and the environmental and ethical questions that follow—Yttrium and Scandium offer a compelling case study. They show that innovation is never just about breakthroughs in labs; it is also about how we mine, refine, recycle, and regulate the elements that power those breakthroughs. And in that understanding lies the blueprint for a future that is not only smarter, but wiser.

**The End**